Weather
Changing Weather and Severe Storms

by Kate Boehm Jerome

Table of Contents

DEVELOP LANGUAGE

Bad **weather** can cause big problems. A **hurricane** is one kind of storm that brings heavy rains, high **winds**, and floods.

People cannot stop a hurricane. But if they know that a hurricane is coming, they can be prepared.

Discuss the photos. Ask and answer questions like these:

What does the large photo show about hurricanes?

Why do you think the people are putting wood over the windows?

Why do you think so many people are driving away from the ocean?

What other things can people do to prepare for a hurricane? Explain your answer.

weather – what is happening in the air around you; the properties of the air at a certain time or place

hurricane – a huge, tropical storm with winds of 119 kilometers (74 miles) per hour or more

winds – the movement of air

hurricane in
Gulf of Mexico
evacuation underway

CHAPTER 1

Describing Weather

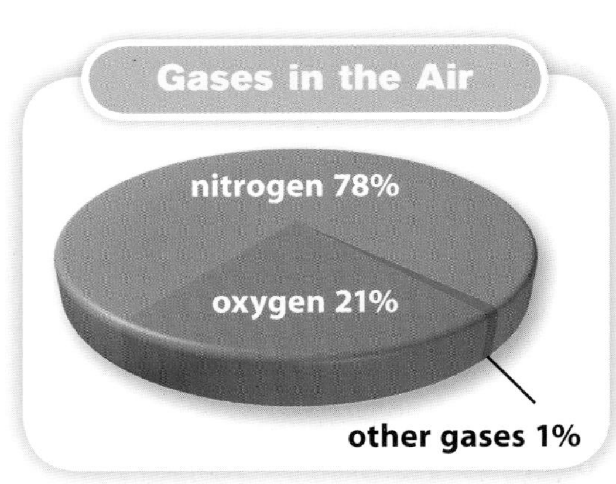

Is the weather hot or cold today? Is the wind blowing? The words you use to describe weather tell what is happening in the air around you.

The air is part of Earth's **atmosphere**. The atmosphere protects Earth by blocking harmful energy from space. The atmosphere also keeps Earth from becoming too hot or too cold. The atmosphere that is closest to Earth contains the gases that most living things need to survive.

Earth's gravity holds the atmosphere in place. But the farther you go away from Earth, the thinner the atmosphere becomes. At about 1,000 kilometers (625 miles) from Earth's surface, the atmosphere becomes so thin that it just disappears into space.

atmosphere – layers of gases that surround Earth

Air Pressure

Scientists measure the **properties** of air to tell us more about the weather. One of the properties they measure is **air pressure**.

Even though you cannot see them, the tiny particles that make up air have weight. The weight of the air is always pressing down on Earth's surface. This creates air pressure.

Air pressure is not the same at all levels of the atmosphere. The particles that make up air are more dense, or more closely packed together, near Earth's surface. Dense air weighs more. So air pressure is higher at sea level than on top of a mountain.

properties – qualities that can be measured or observed

air pressure – the force put on Earth's surface by the weight of the air above it

By The Way...

Have you ever felt a popping in your ears when you are in an elevator or airplane? This happens because the air pressure on the inside of your ears is changing to match the air pressure on the outside of your ears.

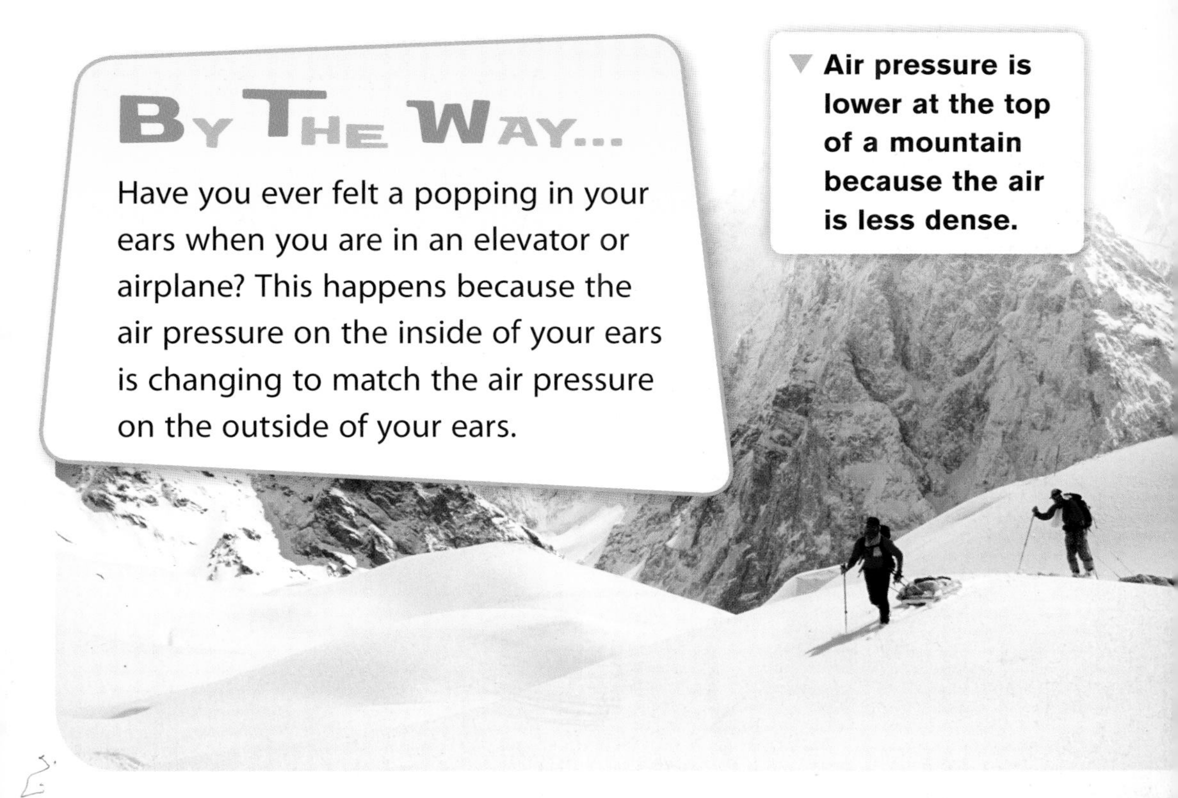

▼ **Air pressure is lower at the top of a mountain because the air is less dense.**

Wind

Air pressure can change even within the same level of the atmosphere. For example, when air gets warm, it rises. The rising particles move farther apart. This makes the air less dense. Since there is less air pressure on Earth underneath the warm air, an area of low pressure forms.

Cold air does the opposite. It gets heavier and sinks. The particles in the cold air move closer together. This makes the air more dense. Since there is more air pressure on Earth underneath the cold air, an area of high pressure forms.

Air usually moves from an area of high pressure to an area of low pressure. This movement of air is wind.

Air moves fastest when there are big differences between pressure areas. Faster moving air makes stronger winds.

Temperature

Another important property of air is **temperature**, or how warm the air is. So how does the air get warm?

Light energy from the sun travels through space and the atmosphere and hits Earth's surface. This light energy warms Earth's surface. Then heat from Earth's surface warms the air above it.

Some places on Earth's surface warm up more than other places. The uneven heating of Earth causes differences in air pressure. Differences in air pressure cause wind.

temperature – a measure of how hot or cold something is

Explore Language

un- = not
uneven = not even, not regular

▼ **Heat from Earth's surface causes the air temperature to increase.**

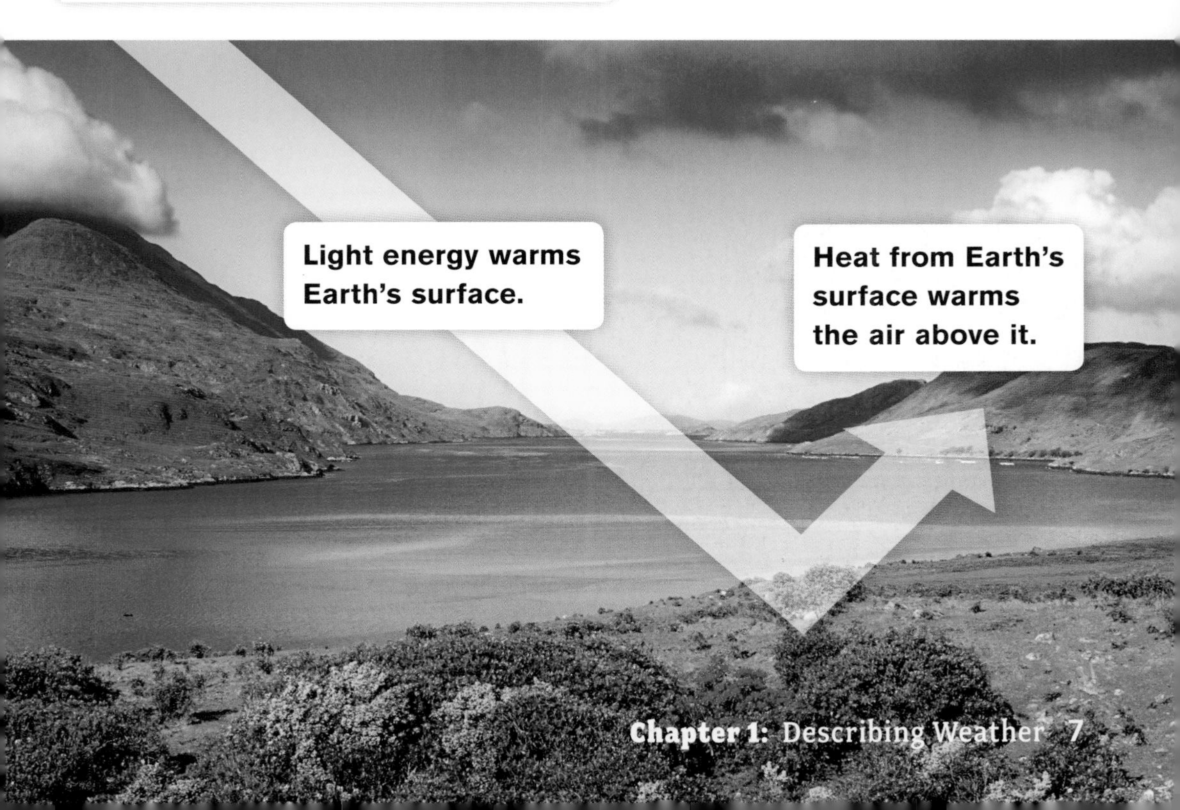

Water Vapor and Humidity

You cannot see it, but there is a lot of water in the air. Water in the air is a gas called **water vapor**. When liquid water heats up enough to change into water vapor, **evaporation** occurs.

Water vapor can change back into a liquid if it cools. When this happens, **condensation** has occurred.

Humidity measures the amount of water vapor in the air. Humidity is another property of air that is used to describe weather. When the air feels damp, or almost wet, we say there is a lot of humidity in the air.

water vapor – water that is in the air as a gas

evaporation – the process by which liquid water on Earth's surface is changed into water vapor

condensation – the process by which water vapor in the air is changed into liquid water

humidity – the amount of water vapor in the air

Water droplets form on this spider web because of condensation.

Evaporation from the ocean adds a lot of water vapor to the air.

KEY IDEAS The atmosphere allows life on Earth to exist. Properties such as air pressure, wind, temperature, and humidity are used to describe weather.

YOUR TURN

OBSERVE

Look at the photos. Use the properties of air to describe the weather in each photo. Write your descriptions in a chart like the one below.

Photo	Description of the Weather
A	
B	

MAKE CONNECTIONS

Air over land warms more quickly than air over water. Explain why this can make the beach a windy place.

USE THE LANGUAGE OF SCIENCE

What are some properties of air that are used to describe weather?

Air pressure, temperature, wind, and humidity are properties that describe weather.

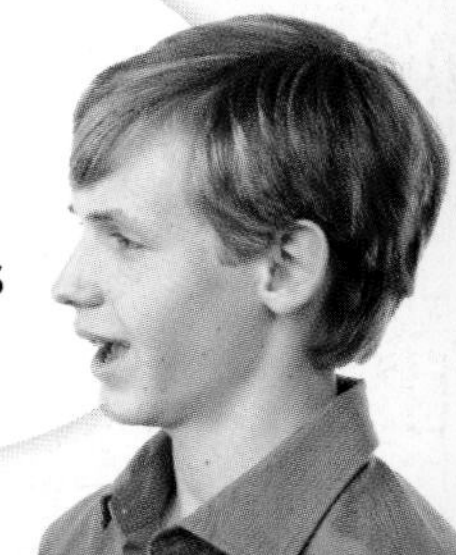

CHAPTER 2

Changes in Weather

Are there many **clouds** in the sky today? If the answer is yes, there might be a change in the weather on the way.

Clouds contain millions of drops of water vapor and **ice crystals**. They form through condensation. There are many different clouds, but all clouds are grouped into three basic kinds. The photos on this page show an example of each kind.

clouds – condensed water vapor in the air

ice crystals – tiny particles of frozen water

▲ Cirrus clouds are high, thin clouds made of ice crystals. They often mean a change in the weather is coming.

▲ Cumulus clouds are puffy, high clouds. They often mean that the weather will be fair.

▲ Stratus clouds form in layers close to the ground. They often bring rain or snow.

Some clouds produce **precipitation**, such as rain or snow. How does this happen? If the water vapor and ice crystals in a cloud become large enough, they will be too heavy to stay in the air. Precipitation occurs when water falls to Earth.

Water moves from place to place through evaporation, condensation, and precipitation. Evaporation moves water into the air. Condensation forms clouds. Precipitation moves water back to Earth's surface.

All of these processes make up the **water cycle**. The water cycle moves water from Earth's surface into the atmosphere and back again. The never-ending water cycle is one reason why the weather is always changing.

precipitation – any form of water that falls from the sky

water cycle – the never-ending pattern of evaporation, condensation, and precipitation that moves water from Earth's surface through the atmosphere and back to Earth's surface again

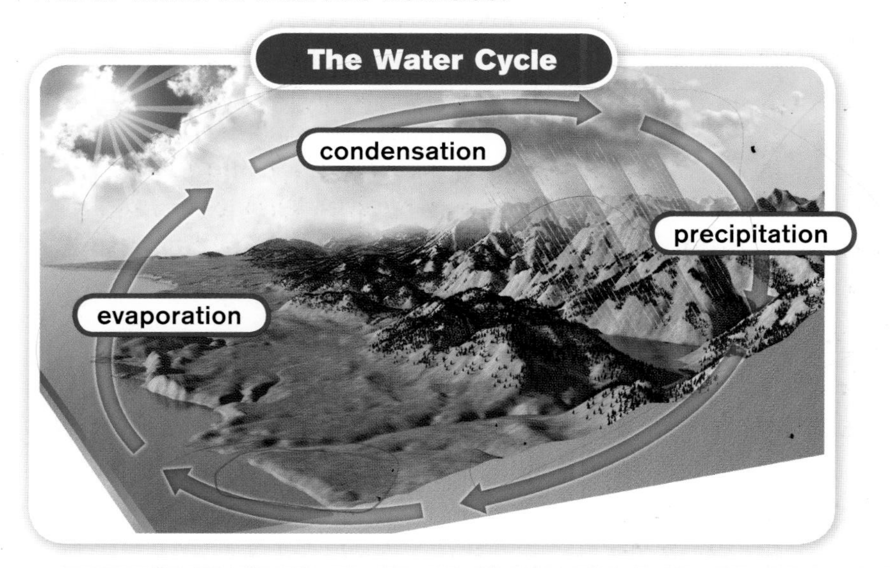

KEY IDEA The processes of evaporation, condensation, and precipitation keep water moving through the water cycle.

Air Masses Bring Weather

Sometimes a line of clouds in the sky can mean that a new **air mass** is moving into an area. An air mass is a large body of air that has almost the same temperature and water vapor levels throughout.

Air masses form over different regions of the world. They take on properties of the air where they are formed.

For example, an air mass that forms over a warm ocean region will have warm, moist air in it. An air mass that forms over a cold northern area will have cold, dry air in it.

air mass – a large body of air with almost the same temperature and water vapor levels throughout

▼ **A cold air mass often forms in Canada and moves down over the United States.**

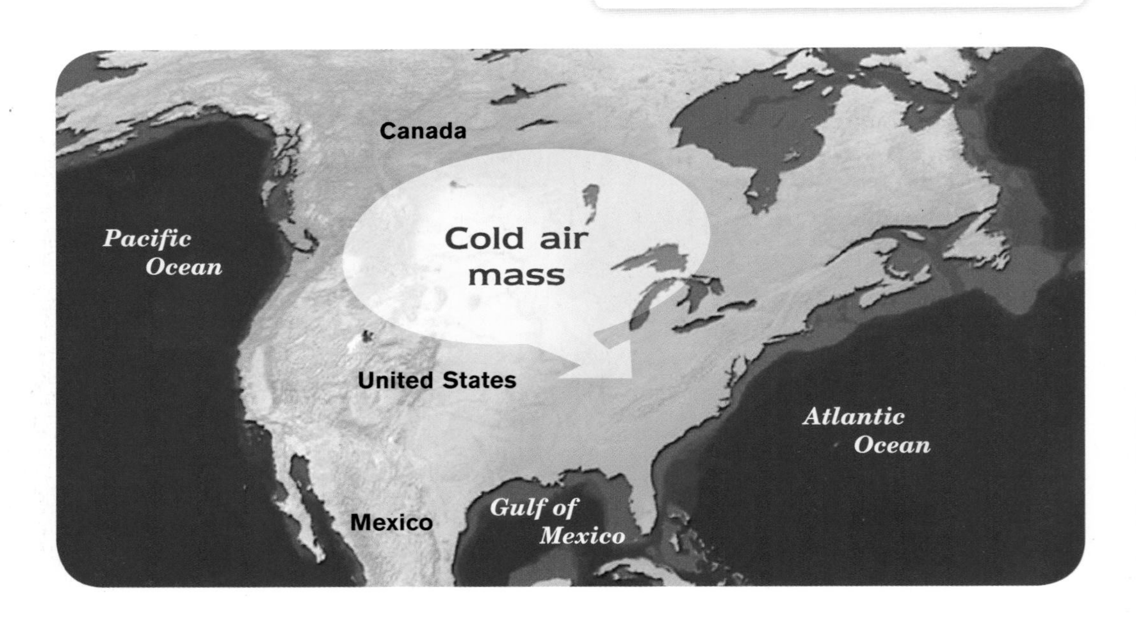

SHARE IDEAS **Describe** the properties of an air mass that forms over the Gulf of Mexico.

Air masses are moved by winds all around Earth. But air masses don't mix when they meet. Instead, a narrow boundary called a **front** is formed.

If cold air pushes into warm air, a cold front is formed. Cold air is more dense so it quickly forces warm air to rise out of the way. If the warm air contains a lot of water vapor, thick clouds can form causing heavy rain and bad weather.

If warm air pushes into cold air, a warm front forms. Warm air slides up and over cold air. This can form high clouds. But as more warm air comes in behind the front, lower clouds can also form.

Precipitation and wind often occur along or near fronts. After a front passes through an area, the weather will have changed.

front – a boundary between air masses

KEY IDEA Air masses move around Earth and bring changing weather with them.

Eyes in the Sky

Weather scientists keep track of fronts with the help of technology, such as weather **satellites**. These satellites orbit Earth above the layers of the atmosphere. They collect images of the weather that is happening in the **troposphere**, the layer of the atmosphere closest to Earth.

The images and other information from weather satellites help scientists track clouds and storms around the world. This helps them predict where the weather will change.

satellites – objects that orbit Earth to collect and send information

troposphere – the layer of the atmosphere closest to Earth's surface

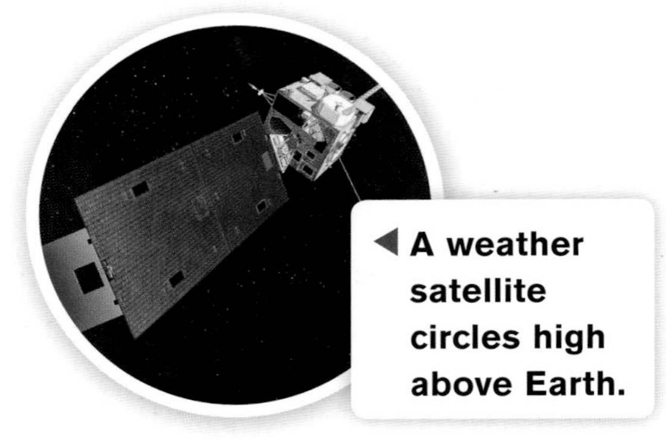

◀ A weather satellite circles high above Earth.

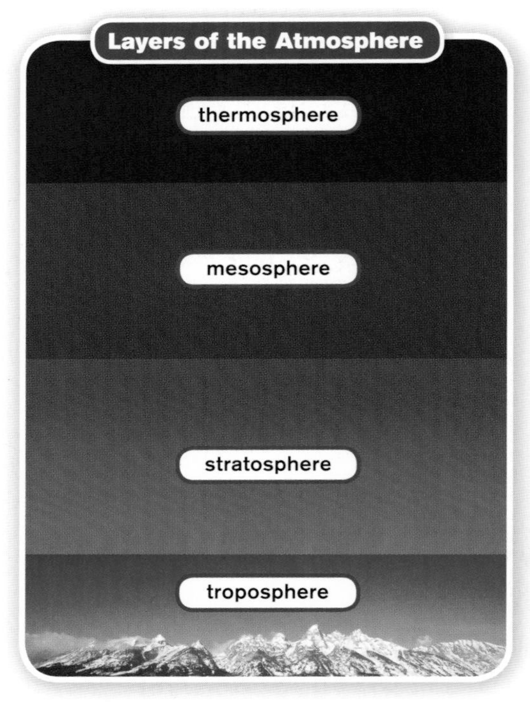

◀ Information from weather satellites shows clouds and other conditions affecting Earth.

KEY IDEA Weather satellites help track weather in the troposphere.

YOUR TURN

PREDICT

A large mass of cold air is moving into an area that has a large mass of warm, humid air. What predictions can you make?

1. Predict what the weather will be like at the front between these two air masses.
2. Predict what the weather will be like after the front moves through the area.

MAKE CONNECTIONS

Would you want to have a picnic when a front is moving through your area? Tell why or why not.

STRATEGY FOCUS

Visualize

Visualize what happens to one drop of water as it moves through the water cycle. Describe what happens at each step.

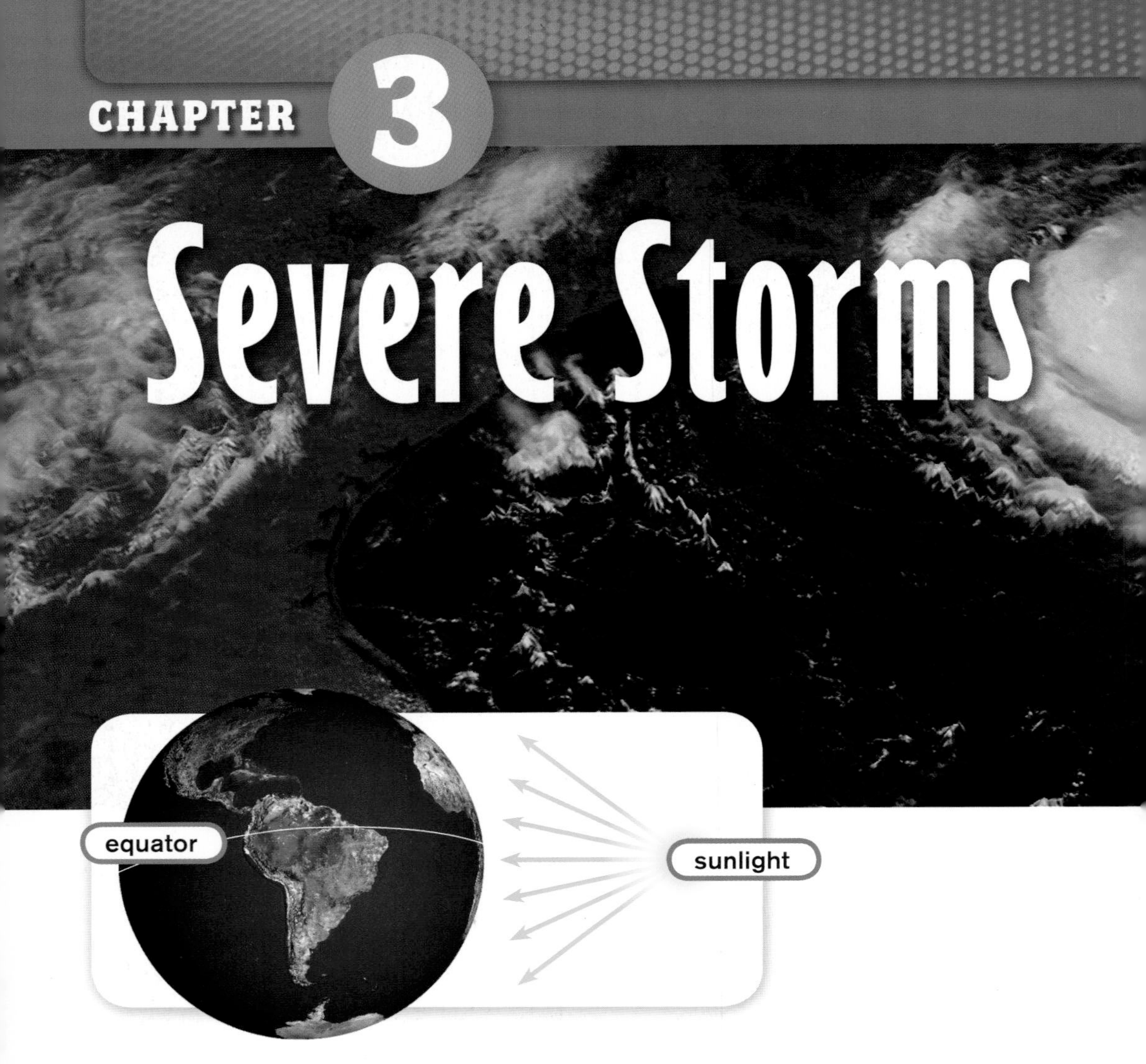

Over a long period of time, the average weather conditions in a certain area form a pattern. This long-term weather pattern is called the area's **climate**.

Since Earth is curved, sunlight does not hit Earth the same in all places.

The most direct sunlight hits Earth at the equator. Since direct sunlight is the strongest, areas around the equator, called the tropics, have the warmest climates.

climate – the long-term weather pattern of an area

Hurricane Katrina in the Gulf of Mexico in 2005

Many people enjoy living in the tropics because of the pleasant weather. But the warm climates also provide conditions for the most powerful storms on Earth. These storms are hurricanes.

Hurricanes can be hundreds of kilometers wide. They form over warm ocean water. Hurricanes that affect the southeastern part of the United States usually occur between June and November.

The very low pressure of a hurricane allows ocean water to rise up in it. As the hurricane moves onto land, it carries lots of water with it. This can lead to severe flooding. Strong winds and rain can also cause a lot of damage.

Since hurricanes are such large storms, they usually move slowly. This can give people time to prepare.

Tornadoes are another form of severe weather. These storms are most likely to form when a dry, cold air mass meets a warm, moist air mass. This can cause violent thunderstorms that produce heavy rains and strong winds.

If conditions are just right, the winds in a thunderstorm can begin to twist. A cone-shaped cloud of whirling wind, called a funnel, forms. If the funnel touches the ground, it is called a tornado. Although a tornado is not as big as a hurricane, its winds can be even stronger. Since tornadoes form so quickly, people often do not have much warning that the storm is coming. Weather scientists are always working on ways to better predict when and where these storms will occur.

tornadoes – funnels of strong, whirling winds that touch the ground and can cause severe damage

This tornado caused damage in Brisco County, Texas, in 2007.

KEY IDEA Hurricanes and tornadoes are severe storms that can cause great damage.

YOUR TURN

INFER

Weather scientists figure out a tornado's wind speed according to how much damage it does.

Look at these two pictures of tornado damage. Infer which tornado had higher wind speeds.

Give details to explain your inference.

MAKE CONNECTIONS

Many of the coastal areas along the southeastern United States and the Gulf of Mexico are only about 3 meters (10 feet) above sea level. Why do you think this might be a problem in a hurricane?

EXPAND VOCABULARY

The suffix *-sphere* means "round." **Atmosphere** and **troposphere** are used in this book. Find out about these other words.

hemisphere **thermosphere** **bathysphere**

What do these words have to do with roundness? Explain what the prefix of each word means, and how the prefix and suffix indicate the meaning of the word.

Meteorologists: What Do They Do?

Meteorologists are scientists who study conditions in the atmosphere. Some meteorologists predict and report the weather on TV. Others do research into such topics as air pollution and future climate changes. But no matter what job they do, all meteorologists love to watch what is happening in the sky!

Places a meteorologist might work	• television station • university • research laboratory • government agency
Education a meteorologist needs	• college degree
Courses you should study if you want to be a meteorologist	• physical science • math • chemistry

This meteorologist at the National Hurricane Center is working to predict when and where a hurricane will reach land.

Question and Answer Words

Sometimes writers ask questions and then answer them. This is one way to share information with readers. The answer often repeats some of the words in the question.

EXAMPLE Example

Why **is** the water cycle **important**?

It **is important** because it provides the water that living things need to survive.

Sometimes writers ask questions and don't answer them. This lets readers think more about the topic on their own.

With a friend, look through this book to find questions in the text. For each question, decide whether or not the answer is in the text. Take turns answering the questions.

Write Questions and Answers

Think about a place you have always wanted to visit.
Find the five-day weather forecast for that place.
Write questions and answers about the weather forecast.

- Use the Internet, a newspaper, or other resources to find the five-day forecast.
- Write questions and answers that give detailed information about the forecast.
- Include a map, a picture, or a chart.

Words You Can Use

Question Words	Weather Words
what	air pressure
how	wind
why	temperature
when	precipitation
where	humidity

Reading a Weather Map

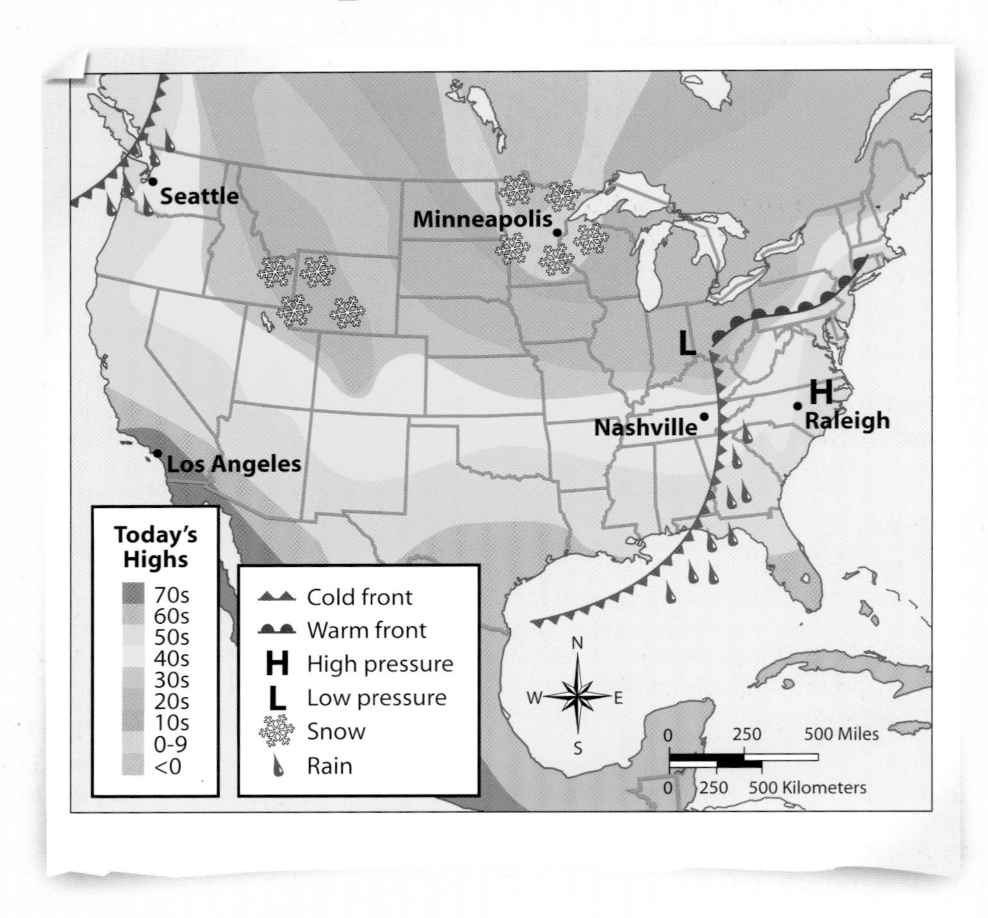

Reading a weather map is easy when you know the basics about weather. Look at the map to answer the questions.

- What city is getting rain?
- What will the temperature be in Los Angeles?
- Which city is getting snow?
- Which city is under an area of high pressure?

Key Words

air mass (air masses) a large body of air with almost the same temperature and water vapor levels throughout
A line of clouds can mean a new **air mass** is moving into an area.

air pressure the force put on Earth's surface by the weight of the air above it
When warm air rises, there is a drop in **air pressure** beneath it.

atmosphere layers of gases that surround Earth
The **atmosphere** protects life on Earth.

climate (climates) the long-term weather pattern of an area
The tropics have very warm **climates**.

condensation the process by which water vapor in the air is changed into liquid water
The **condensation** of water vapor forms clouds in the sky.

evaporation the process by which liquid water on Earth's surface is changed into water vapor
Water changes from a liquid to a gas during **evaporation**.

front (fronts) a boundary between air masses
When a cold air mass pushes into a warm air mass, a cold **front** is formed.

humidity the amount of water vapor in the air
A high **humidity** level can make the air feel damp, or almost wet.

precipitation any form of water that falls from the sky
In northern climates, snow is a common form of winter **precipitation**.

temperature (temperatures) a measure of how hot or cold something is
The **temperature** of the air rises when Earth's surface beneath it gets warmer.

troposphere the layer of the atmosphere closest to Earth's surface
Weather satellites collect images of weather happening in the **troposphere**.

water cycle the never-ending pattern of evaporation, condensation, and precipitation that moves water from Earth's surface through the atmosphere and back to Earth's surface again
Clouds are part of the **water cycle**.

water vapor water that is in the air as a gas
Liquid water that turns into a gas is called **water vapor**.

weather what is happening in the air around you; the properties of the air at a certain time or place
Weather can be rainy one day and sunny the next.

wind (winds) the movement of air
A tornado has very strong **winds**.

Index

MILLMARK EDUCATION CORPORATION
Ericka Markman, President and CEO; Karen Peratt, VP, Editorial Director; Lisa Bingen, VP, Marketing; Rachel L. Moir, Director, Operations and Production; Shelby Alinksy, Assistant Editor; Mary Ann Mortellaro, Science Editor; Pictures Unlimited, Photo Research

PROGRAM AUTHORS
Mary Hawley; Program Author, Instructional Design
Kate Boehm Jerome; Program Author, Science

BOOK DESIGN Steve Curtis Design

CONTENT REVIEWER
Tom Nolan, Operations Engineer, NASA Jet Propulsion Laboratory, Pasadena, CA

PROGRAM ADVISORS
Scott K. Baker, PhD, Pacific Institutes for Research, Eugene, OR
Carla C. Johnson, EdD, University of Toledo, Toledo, OH
Donna Ogle, EdD, National-Louis University, Chicago, IL
Betty Ansin Smallwood, PhD, Center for Applied Linguistics, Washington, DC
Gail Thompson, PhD, Claremont Graduate University, Claremont, CA
Emma Violand-Sánchez, EdD, Arlington Public Schools, Arlington, VA (retired)

TECHNOLOGY
Arleen Nakama, Project Manager
Audio CDs: Heartworks International, Inc.
CD-ROMs: Cannery Agency

PHOTO CREDITS cover ©Weatherpix; IFC and 15b ©David Safanda/iStockphoto.com; 1 ©tomasovic_net/Shutterstock; 2-3 ©Jeff Schmaltz/MODIS Rapid Reponse Team, NASA/GSFC; 3a ©Kin Man Hui/SanAntonioExpress/ZUMA Press; 3b ©Chris Carson/UPI Newspictures; 3c ©U.S. Navy/ZUMA Press; 4a illustration by Joel and Sharon Harris; 4b ©Tom and Pat Leeson Photography; 5 ©Colin Monteath/agefotostock; 6a ©Terry Qing/Getty Images; 6b, 11, 12, 13a, 13b illustrations by Rob Kemp; 7 ©The Irish Image Collection/agefotostock; 8a ©Marika Eglite/Shutterstock; 8b ©Momatiuk-Eastcott/agefotostock; 9a ©Amanda Voisard/ZUMA Press; 9b ©Motofish Images/Corbis; 9d and 9e Ken Karp for Millmark Education; 10a ©Adam Jones/Visuals Unlimited; 10b ©iofoto/Shutterstock; 10c ©David R. Frazier/Photo Researchers, Inc.; 14a, 14b, 16-17 ©NOAA; 14c ©Stockbyte/Punchstock; 15a ©Tony Freeman/PhotoEdit; 16a ©MSA/Photo Researchers; 18 and 24 ©Reed Timmer/Jim Reed Photography/Corbis; 19a ©John Terhune/Journal and Courier/AP images; 19b ©Larry W. Smith/epa/Corbis; 20 ©Getty Images; 22 map by Mapping Specialists

Published by Millmark Education Corporation
7272 Wisconsin Avenue, Suite 300
Bethesda, MD 20814

ISBN-13: 978-1-4334-0203-6

Printed in the USA

10 9 8 7 6 5 4 3 2 1